MENTAL MATHS

Book 4

C.V. Indira

BPI INDIA PVT LTD

© BPI INDIA PVT LTD, 2007

First revised edition: 2010

Reprinted: 2008, 2009, 2011

All rights reserved. No part of this publication may be reproduced or transmitted, in any form or by any means, without permission. Any person who does any unauthorised act in relation to this publication may be liable to criminal prosecution and civil claims for damages.

ISBN: 978817693149-6

The views expressed and the material provided in this book are solely those of the author and presented by the publisher in good faith. The publisher is in no way responsible for the same.

C.V. Indira

BPI INDIA PVT LTD

F-213/A, Ground Floor, Old Mehrauli Badarpur Road,
Lado Sarai, New Delhi- 110030 (India)
Tel: +91-11-43394300-99
E-mail: sales@bpiindia.com

Buy online
www.bpiindia.com

Contents

1. Numbers .... 5 - 10

2. Factors and Multiples .... 11 - 14

3. Fractions .... 15 - 18

4. Addition .... 19 - 22

5. Subtraction .... 23 - 26

6. Multiplication .... 27 - 33

7. Division .... 34 - 36

8. Measurement .... 37 - 40

9. Money .... 41 - 43

10. Time .... 44 - 56

1 Numbers

 REVISION

Write the following in words.

i) 335

ii) 447

iii) 798

iv) 321

v) 129

Write the following in numerals.

i) Seven hundred and forty nine

ii) Two hundred and eleven

iii) Four hundred and nine

iv) Six hundred and sixteen

v) Five hundred and thirty five

Write in words.

i) 3719

ii) 6897

iii) 6009

iv) 5329

v) 1297

vi) 7895

vii) 5309

viii) 4849

FIVE AND SIX DIGIT NUMBERS

10000 is the smallest 5 digit number

100000 is the smallest 6 digit number

Place value

Ten thousand	Thousand	Hundred	Tens	Units
10000	1000	100	10	1

Example

Place 37482 in the place value table

Ten thousand	Thousand	Hundred	Tens	Units
3	7	4	8	2

 EXERCISE

1. Arrange the following in a place value table.

i) 3140   ii) 377   iii) 4321   iv) 6285   v) 51354

vi) 67239

2. Write the following in words.

i) 2985   ii) 37895   iii) 4590   iv) 5987

v) 29850   vi) 19599

3. Which is the highest 5-digit number?

4. Which is the highest 6-digit number?

5. Write the place value of the following.

i) 9 in 2985   ii) 6 in 68958   iii) 8 in 98116

iv) 7 in 75950   v) 4 in 958490

6. Write the digit with the greatest place value in the following.

i) 8758   ii) 575038   iii) 389   iv) 63584

EXPANDED NOTATION

Example

Write the expanded form of 3978

Expanded form of 3978 = 3000 + 900 + 70 + 8

 EXERCISES

1. Write the expanded form of the following.

i) 2985   ii) 3793   iii) 950389

iv) 1935   v) 109538

2. Write the numeral for the following.

i) 30000 + 4000 + 300 + 20 + 1

ii) 70000 + 3000 + 500 + 40 + 3

iii) 90000 + 7000 + 400 + 30 + 9

iv) 8000 + 600 + 30 + 9

v) 900 + 60 +3

3. Which is the largest number?

i) 7326   7658   8932   9853

ii) 8354   8453   8534   8435

iii) 98346   83946   94638   96348

iv) 7312   7213   7123   7231

4. Arrange the following in ascending order.

i) 3912, 4314, 9813, 2252, 6896

ii) 94321, 65428, 73124, 12535

iii) 7532, 8959, 4754, 3213

iv) 4371, 9859, 2341, 7598

5. Arrange the following in descending order.

i) 7959, 3894, 8758, 2959

ii) 385, 983, 777, 895

iii) 8757, 6896, 7598, 8953

iv) 395, 837, 285, 594

GREATEST AND SMALLEST NUMBERS

Write the smallest 4-digit number and the greatest 4-digit number with the following

3, 8, 5, 2

For the greatest 4-digit number we place the smallest digit in ones' place, next greater in tens' place and so on i.e. 8532 is the greatest 4-digit number that can be formed with the above digits.

Similarly, for the smallest we write the greatest digit in the ones' place, the next smaller in the tens' place and so on i.e. 2358 is the smallest 4-digit number that can be formed with the above digits.

 EXERCISES

1) Write the greatest and the smallest 3-digit numbers (without repeating the digits).

   i) 3, 8, 7   ii) 7, 5, 9   iii) 4, 8, 3   iv) 5, 2, 1   v) 7, 8, 0

2) Write the greatest and the smallest 4-digit numbers (without repeating the digits).

   i) 4, 9, 7, 2   ii) 3, 8, 9, 5   iii) 5, 6, 7, 9

   iv) 7, 5, 9, 2   v) 6, 1, 7, 3

3) Write the greatest and the smallest 5-digit numbers (without repeating the digits).

   i) 7, 3, 2, 1, 5   ii) 2, 9, 4, 6, 1   iii) 5, 9, 4, 2, 8

   iv) 8, 3, 1, 6, 5   v) 4, 1, 2, 3, 5

4) Write the greatest 5-digit numbers (digits may be repeated).

    i) 2, 5, 3, 4, 9   ii) 7, 9, 2, 3, 1   iii) 9, 8, 5, 7, 6

    iv) 3, 2, 1, 4, 5   v) 1, 9, 8, 7, 6

ROUNDING NUMBERS

Examples

i) Round 38 to the nearest 10.

Observe that 38 is nearer to 40 than it is to 30. Hence 38 is rounded to 40.

ii) Round 293 to the nearest 100.

Observe here that 293 is closer to 300 than it is to 200. Hence 293 is rounded to 300.

 EXERCISES

1. Round the following numbers to the nearest 10.

    i) 62   ii) 838   iii) 57   iv) 249   v) 437

2. Round the following to the nearest 100.

    i) 211   ii) 385   iii) 6119   iv) 7538   v) 61015

3. Round the following to the nearest 1000.

    i) 9208   ii) 6921   iii) 1238   iv) 45932   v) 98210

2 Factors and Multiples

FACTORS

Examples

$$\begin{array}{r} 5 \\ 4\overline{)20} \\ \underline{20} \\ 0 \end{array}$$

4 divides 20 leaving no remainder. Hence 4 is a factor of 20.

$$\begin{array}{r} 2 \\ 8\overline{)20} \\ \underline{16} \\ 4 \end{array}$$

8 does not divide 20 completely, hence 8 is not a factor of 20.

Observe

$$\begin{array}{r} 4 \\ 5\overline{)20} \\ \underline{20} \\ 0 \end{array} \quad \begin{array}{r} 5 \\ 4\overline{)20} \\ \underline{20} \\ 0 \end{array} \quad \begin{array}{r} 20 \\ 1\overline{)20} \\ \underline{20} \\ 0 \end{array} \quad \begin{array}{r} 1 \\ 20\overline{)20} \\ \underline{20} \\ 0 \end{array}$$

From the above example we see that 4, 5, 20, 1 are factors of 20.

When two numbers are multiplied, each number in the product is a factor of that product.

Find out if 9 is a product of 32.

```
      3
9 ) 32
    27
   ---
     5
```

Since 9 does not divide 32 completely, it is not a factor of 32.

 EXERCISES

1) Find the factors of

   16, 32, 18, 25

2) Find out whether

i) 7 is a factor of 27

ii) 4 is a factor of 24

iii) 10 is a factor of 45

3) Find the least 2-digit number that has

i) 3 and 4 as factors

ii) 2, 5, 8 as factors

MULTIPLES

If 5 is a factor of 10 then 10 is a multiple of 5.

From this we can conclude that factors and multiples are related to each other.

Examples

1) 4 × 8 = 32

Here 32 is a multiple of 4 and 8.

2) 3 × 9 = 27

Here 27 is a multiple of 3 and 9.

3) Find out if 76 is a multiple of 9.

```
     8
9 ) 76
    72
    ─
     4
```

Since 9 does not divide 76 completely 9 is not a factor of 76. And therefore, 76 is not a multiple of 9.

We can distinguish between an even and odd number using factors and multiples, i.e. an even number has 2 as a factor and is also a multiple of 2.

 EXERCISES

1. Write 4 multiples of

i) 3

ii) 6

iii) 7

iv) 5

v) 8

vi) 11

2. Find out if

i) 36 is a multiple of 9

ii) 25 is a multiple of 5

iii) 120 is a multiple of 8

3. Find out whether odd or even.

14, 53, 26, 46, 58, 27, 19, 18, 25, 38, 36, 42

4. Write all odd numbers between.

a) 23 and 33

b) 10 and 25

3 Fractions

REVISION

1. Write the following fractions in words.

i) 3/7   ii) 2/5   iii) 1/4   iv) 9/10

Equal fractions

If two fractions represent the same portion of a whole, they are said to be equal or equivalent fractions.

Example

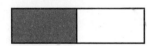

$\dfrac{1}{2}$

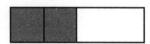

$\dfrac{2}{4}$

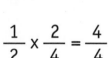

$\dfrac{1}{2} \times \dfrac{2}{4} = \dfrac{4}{4}$

When cross-multiply two equivalent fractions, we get the same number as numerator and denominator.

EXERCISES

1. Find out if two fractions are equal or not.

a) $\dfrac{3}{7}$, $\dfrac{4}{8}$ b) $\dfrac{2}{5}$, $\dfrac{4}{16}$ c) $\dfrac{1}{4}$, $\dfrac{6}{24}$

d) $\dfrac{2}{7}, \dfrac{3}{7}$ e) $\dfrac{5}{10}, \dfrac{18}{36}$

PROPER AND IMPROPER FRACTIONS

Proper fractions are those fractions in which the numerator is less than the denominator. Those fractions in which the numerator is equal to unity (1) are called unit fractions.

e.g. $\dfrac{1}{2}, \dfrac{1}{8}, \dfrac{1}{5}, \dfrac{1}{3}$

Improper fractions are those fractions in which the numerator is greater than the denominator.

A mixed number is made up of a whole number and a proper fraction. A mixed number can be converted into an improper fraction and an improper fraction can be written as a mixed number.

Examples

1. Write $2\dfrac{3}{4}$ as an improper fraction

$$2\dfrac{3}{4} = \dfrac{2}{1} + \dfrac{3}{4} = \dfrac{2 \times 4 + 3}{4} = \dfrac{8 + 3}{4} = \dfrac{11}{4}$$

2. Write $\dfrac{56}{16}$ as a mixed number

$$\dfrac{56}{16} = \dfrac{14}{4} = \dfrac{7}{2}$$

Denominator 2)7̄ 3 whole number part of the fraction
         6
         1 Numerator $= 3\dfrac{1}{2}$

EXERCISES

1. Write the following as improper fractions.

i) $1\frac{3}{4}$ ii) $3\frac{6}{7}$ iii) $11\frac{3}{5}$ iv) $7\frac{5}{8}$

2. Write as mixed numbers.

i) $\frac{16}{15}$ ii) $\frac{23}{7}$ iii) $\frac{91}{14}$ iv) $\frac{100}{4}$

LIKE FRACTIONS AND UNLIKE FRACTIONS

Fractions, which have the same denominator, are called like fractions, e.g. $\frac{1}{3}$ and $\frac{2}{3}$ are like fractions. Fractions which have different denominators are called unlike fractions, e.g. $\frac{7}{5}$ and $\frac{1}{3}$ are unlike fractions.

 EXERCISES

1. Pick out the like fractions from the following.

i) $\frac{4}{7}$, $\frac{3}{7}$ and $\frac{5}{7}$ ii) $\frac{1}{3}$, $\frac{1}{2}$ iii) $\frac{5}{4}$, $\frac{3}{4}$

iv) $\frac{1}{4}$, $\frac{5}{4}$ v) $\frac{3}{2}$, $\frac{7}{4}$

ADDITION AND SUBTRACTION OF LIKE FRACTIONS

 **REVISION**

1. Find the sum.

i) $\dfrac{2}{7} + \dfrac{3}{7}$ ii) $\dfrac{1}{5} + \dfrac{2}{5}$ iii) $\dfrac{1}{4} + \dfrac{3}{4}$

iv) $\dfrac{1}{9} + \dfrac{5}{9}$ v) $\dfrac{3}{8} + \dfrac{2}{8}$

2. Find the difference.

a) $\dfrac{3}{7} - \dfrac{2}{7}$ b) $\dfrac{4}{6} - \dfrac{1}{6}$ c) $\dfrac{8}{10} - \dfrac{6}{10}$

d) $\dfrac{5}{9} - \dfrac{1}{9}$ e) $\dfrac{3}{8} - \dfrac{1}{8}$

ORDERING LIKE FRACTIONS

1. Arrange the following in ascending order.

i) $\dfrac{3}{7}, \dfrac{1}{7}, \dfrac{4}{7}, \dfrac{6}{7}$ ii) $\dfrac{1}{9}, \dfrac{8}{9}, \dfrac{5}{9}, \dfrac{3}{9}, \dfrac{6}{9}$

iii) $\dfrac{3}{11}, \dfrac{8}{11}, \dfrac{9}{11}, \dfrac{4}{11}, \dfrac{6}{11}$

2. Write the correct symbol (> or <).

i) $\dfrac{7}{9}$ ___ $\dfrac{1}{9}$ ii) $\dfrac{5}{4}$ ___ $\dfrac{3}{4}$

iii) $\dfrac{1}{3}$ ___ $\dfrac{2}{3}$ iv) $\dfrac{4}{7}$ ___ $\dfrac{6}{7}$

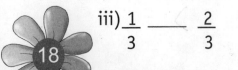

4 Addition

 REVISION

```
    1 1 7 2          3 6 5 7          3 2 7 4
+   1 5 1 7      +   2 3 5 1      +   2 1 0 2
_____        _____        _____


    4 0 1 3          4 2 1 0          3 2 7 5
+   2 2 1 1      +   3 8 8 1      +   2 9 8 7
_____        _____        _____


    5 9 5 4          6 7 3 7          1 9 5 0
+   4 2 3 5      +   3 2 8 1      +   1 0 1 2
_____        _____        _____
```

1. Find the sum.

```
   76543          30904          64458
+  32104       +  49036       +  24416
  _____         _____         _____


   19419          25357          37593
+  80501       +  12212       +  17212
  _____         _____         _____


   49543          12045          74890
+  43537       +  57353       +  26775
+  31987       +  48273       +  96453
  _____         _____         _____


   38259          89797          13325
+  59542       +  52671       +  31987
+  38815       +  48273       +  76803
  _____         _____         _____
```

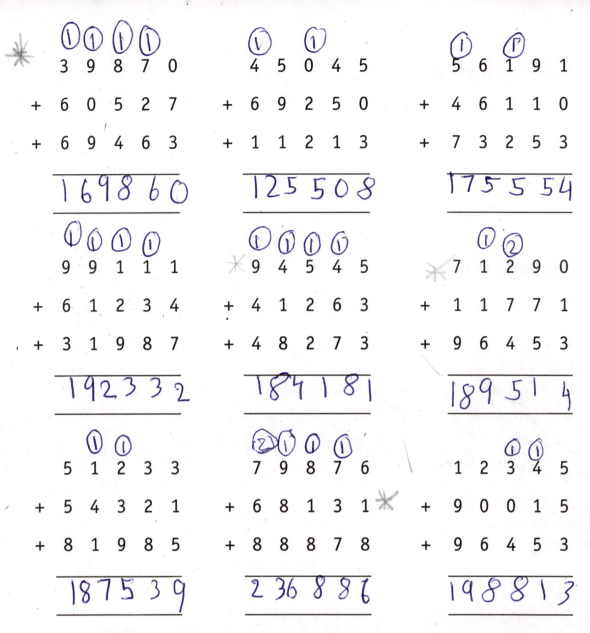

2. Find the missing digits.

i) 4 3 ② 1 5
  + 3 ⑧ 2 9 0
  ─────────
    8 1 5 0 ⑤

i) 2 9 5 ② 0
  + 1 8 ⑨ 5 0
  ─────────
    4 ⑧ 4 7 ⓪

```
iii)   3 8 2 *         iv)   4 2 * 3
     + 2 7 * 5              + * 9 7 5
     + 1 * 7 4              + 2 * 1 *
     ─────────              ─────────
     * 1 * 2                8 5 4 7
     ─────────              ─────────
```

PROBLEMS

1. What number should be added to 57182 to get 91542?

2. What number should be added to 4985 to get 9732?

3. A fruit seller bought 2314 apples from one place, 4312 apples from another and 3313 apples from a third place. How many apples did he buy in all?

4. There are 3 schools in a town. One school has 2312 students, the second one has 4329 and the third school has 3825 students. How many students are there in all?

5. There are three villages near a town. The first village has a population of 59231, out of which 29324 are women. The second village has a population of 32952, out of which 10842 are women and the third has a population of 9342, out of which 2320 are women. How many men are there in the villages in all?

5 Subtraction

 REVISION

```
  7 7 5 4          4 1 1 4          7 8 9 6
- 4 6 5 3        - 3 0 0 4        - 6 8 8 5
  -------          -------          -------
  3 1 0 1          1 1 1 0          1 0 1 1

    6 16                              8 7 7 15
  8 7 6 8          4 0 0 9          9 8 8 5
- 8 4 9 3        - 3 0 0 4        - 4 8 8 7
  -------          -------          -------
  0 2 7 5          1 0 0 5          4 7 7 8

    16                              
  15 7 18          1 10 14            8 18
  7 6 7 8          6 8 1 4          9 8 8 8
- 6 3 8 9        - 2 1 7 7        - 6 8 7 9
  -------          -------          -------
                   4 0 3 7          3 0 1 9

  5 7 16
  6 8 8 16         9 7 8 9          7 9 1 8
- 5 3 7 9        - 5 7 7 4        - 6 7 0 4
  -------          -------          -------
  1 2 2 7          4 0 1 5          1 2 1 4
```

```
   9 0 8 7          6 ⁶7̶ ¹²8̶ 9          ⁹9 ⁷8̶ ⁶7̶ ¹⁰0̶
 - 4 0 5 6        - 5 0 9 9          - 3 5 8 9
 ─────────        ─────────          ─────────
   5 0 ③ 1          1 6 9 0            6 2 1 1
       3
```

 **EXERCISES**

1. Find the difference.

```
   7 5 5 6 3        8 7 3 2 1        2 6 6 5 9
 - 5 2 8 9 4      - 5 9 0 5 6      - 2 4 7 8 9
 ───────────      ───────────      ───────────


   8 0 5 0 1        2 5 8 8 8        3 7 5 9 3
 - 1 9 4 1 9      - 1 7 9 9 9      - 1 7 2 1 2
 ───────────      ───────────      ───────────


   7 6 6 4 3        9 0 9 0 4        6 4 2 4 8
 - 3 8 7 6 4      - 8 9 0 5 6      - 4 4 4 1 6
 ───────────      ───────────      ───────────
```

EXERCISES

1. Estimate the following by rounding to the nearest 10.

i) 35 × 46   ii) 58 × 72   iii) 85 × 39

iv) 28 × 82   v) 69 × 22   vi) 36 × 78

2. Estimate the following to the nearest 10 by rounding the first number up and the second number down.

i) 36 × 44   ii) 28 × 38   iii) 75 × 35

iv) 43 × 22   v) 97 × 79

3. Estimate the following to the nearest 10 by rounding the first number down and the second number up.

i) 87 × 27   ii) 56 × 64   iii) 32 × 43

iv) 92 × 29   v) 33 × 44

4. PROBLEMS

i) There are 25 eggs in a crate. There are 75 such crates in a truck. How many eggs are there in all?

ii) Ramu earns Rs 2550 a month. How much money will he earn in 7 years?

iii) Rajiv has 23 stamp albums. Each album contains 455 stamps. How many stamps are there in all?

iv) A milk van carries 525 packets of milk everyday. How many packets would 19 such vans carry?

7 Division

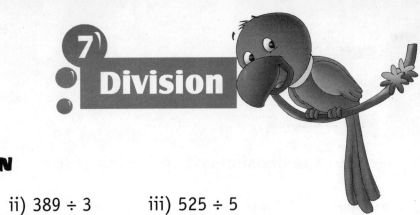

 REVISION

i) 48 ÷ 7   ii) 389 ÷ 3   iii) 525 ÷ 5

iv) 895 ÷ 9   v) 421 ÷ 3

Division of 4-digit numbers by a single digit number.

Example

```
       9 5 2
    4)3808
     - 36
        20
       -20
        08
       -08
         0
```

To verify 952 × 4 = 3808

Example

```
       2477
    3)7432
     - 6
       14
      - 12
        23
       - 21
         22
        - 21
          1
```

To verify

2477 × 3 + 1 = 7431 + 1 = 7432

 EXERCISES

1. Carry out the division and verify the answer.

i) 2)9321   ii) 3)7231   iii) 4)4484
iv) 7)8721   v) 9)3323   vi) 8)7882
vii) 5)4359 viii) 6)6895 ix) 3)2935
x) 7)3439

2. Find the quotient.

i) 45 ÷ 10   ii) 50 ÷ 10   iii) 65 ÷ 10
iv) 270 ÷ 20 v) 235 ÷ 30   vi) 800 ÷ 40
vii) 1139 ÷ 10

ESTIMATION OF QUOTIENT

Example

Estimate the quotient for 64 ÷ 36

Thus 60 ÷ 40 = 6 ÷ 4, which is approximately equal to 1

 EXERCISES

1. Find the estimated quotient for each of the following.

i) 84 ÷ 23   ii) 175 ÷ 25   iii) 633 ÷ 33
iv) 484 ÷ 22 v) 385 ÷ 15

DIVIDING BY A 2-DIGIT NUMBER

Example

Divide 489 by 12

```
        40
   12) 489
        48
        09
```

To verify the answer 12 × 40 + 9 = 480 + 9 = 489

 EXERCISES

1. Carry out the following divisions and verify the answers.

i) 385 ÷ 35   ii) 489 ÷ 63   iii) 543 ÷ 23

iv) 689 ÷ 13   v) 709 ÷ 70

Measurement

LENGTH

 REVISION

1. Find the length of the following line segments.

i) A_____B

ii) A_____B

iii) A _____B

2. Draw a line segment of the following measurement.

i) 4 cm   ii) 7 cm   iii) 3 ½ cm

iv) 6 cm   v) 5 ½ cm

PERIMETER

The distance of the boundary of a closed figure or the total length of the sides of a polygon is called its perimeter.

(A polygon is a closed plane figure formed by three or more line segments that do not cross each other).

Example

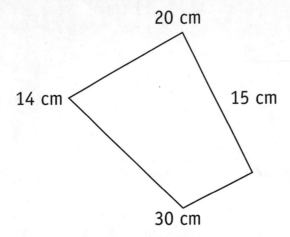

Perimeter = 14 + 20 + 15 + 30 = 34 + 45 = 79 cm

Therefore, the perimeter of the given figure is 79 cm.

 EXERCISES

1. Find the perimeter of the following.

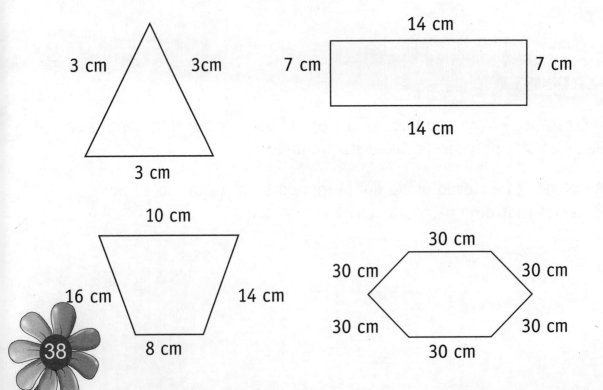

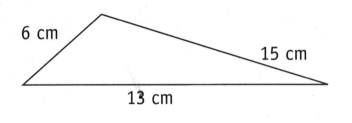 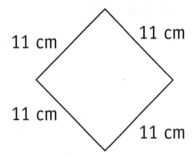

MASS

The units of measurement of mass are gram (g) and kilogramme (kg)

1 kg = 1000 g

1. How do you measure the following - in grams or kilogrammes?

i) A sack of potatoes

ii) A book

iii) A pen

2. If one kilogramme of potatoes costs Rs 15, how much would.

i) 3 kg of potatoes cost?

ii) 5 kg of potatoes cost?

iii) 15 kg of potatoes cost?

iv) 390 kg of potatoes cost?

v) 483 kg of potatoes cost?

CAPACITY

The units of capacity are millilitres (ml) and litres (l).

1 litre = 1000 ml

 **REVISION**

1. How would you measure the following (ml or l)?

i) One cup of water

ii) Petrol in a car

iii) Medicine dosage

9 Money

REMEMBER

1 Rupee = Two 50 paise coins or Four 25 paise coins

4 Rupees 50 paise is written as Rs 4.50

25 Rupees 75 paise is written as Rs 25.75

 EXERCISES

1. How many 50 paise coins will you get from 7 Rupees?

2. A man had a 20 Rupee note and a 50 Rupee note. He spends Rs 65 on groceries. How much money does he have now?

3. What notes and coins does Rs 14.75 have?

Bills

Example

Ravi bought sugar for Rs 15.50, sweets for Rs 35.25 and salt for Rs 6.50. Make a bill for these items.

Item	Amount (Rs)
Sugar	15.50
Sweets	35.25
Salt	6.50

EXERCISES

1. Prepare a bill for the following

i) 3 kg rice (@ Rs 12/kg), 2 kg sugar (@ Rs 16/kg) and 5 litres of oil (@ Rs 55/l).

ii) Mangoes – Rs 15.60, Apples – Rs 13.25, Bananas – Rs 5.75 and Grapes – Rs 25.30.

PROBLEMS

1. Vandana bought two books. The first book cost her Rs 165.50 and the second Rs 153.25. How much did she spend on the books?

2. Out of the money that he had in his pocket father gave Rs 11.75 to Raju. He's now left with Rs 61.25. How much did he have in his pocket initially?

3. 15 packets of peanuts cost Rs 25.50. What is the cost of one packet?

4. Renu needs Rs 535 for her school fete. She has collected Rs 398 so far. How much more money does she need?

5. The newspaper bill for a month is Rs 109.25. What is the cost of one newspaper?

6. Tanay spends Rs 250 per month on his notebooks. How much does he spend per day?

7. The prices of some items in a bakery are given below. Ravi needs to buy 2 packets of biscuits, 2 cakes, 35 pastries and ½ kg of cookies for his birthday. How much would all these items cost him?

Item	Price (Rs)
Biscuits	20.25 per packet
Cakes	13.00 per cake
Pastries	5.25 per piece
Cookies	92.25 per kg

10 Time

FACTS

60 seconds = 1 minute

60 minutes = 1 hour

24 hours = 1 day

7 days = 1 week

365 days = 1 year

12 months = 1 year

52 weeks = 1 year

10 years = 1 decade

100 years = 1 century

1000 years = 1 millennium

The month of February has only 28 days. But once in every 4 years February has 29 days and that year (leap year) has 366 days.

Conversion of weeks into days, days into weeks

Example 1

Convert 14 weeks into days

1 week = 7 days

14 weeks = 14 × 7 = 98 days

Example 2

Convert 42 days into weeks

7 days = 1 week

42 days = 42 ÷ 7 = 6 weeks

Example 3

Convert 55 days into weeks

7 days = 1 week

55 days = 55 ÷ 7

```
      7
   7)55
   - 49
      6
```

There are 7 weeks in 55 days and the remainder 6 is the number of days.

 EXERCISES

1. Convert to days.

i) 14 weeks   ii) 27 weeks   iii) 67 weeks

iv) 45 weeks   v) 54 weeks   vi) 78 weeks

vii) 93 weeks   viii) 17 weeks   ix) 29 weeks

x) 77 weeks

2. Convert to weeks.

i) 120 days   ii) 85 days   iii) 90 days

iv) 123 days   v) 200 days   vi) 156 days

vii) 67 days   viii) 365 days   ix) 203 days

x) 84 days

Hours to minutes and seconds, and minutes to seconds

Example 1

Convert 5 hours into minutes

1 hour = 60 minutes

5 hours = 5 × 60 minutes

= 300 minutes

Example 2

Convert 5 hours 25 minutes into minutes

1 hour = 60 minutes

5 hours = 5 × 60 minutes

5 hours and 25 minutes = 300 + 25 minutes

= 325 minutes

Example 3

Convert 120 minutes into hours
60 minutes = 1 hour
120 minutes = 120 ÷ 60
= 2 hours

```
       2
60)120
   -120
      0
```

Example 4

Convert 20 minutes into seconds

1 minute = 60 seconds

20 minutes = 20 × 60 =1200 seconds

Example 5

Convert 360 seconds into minutes
60 seconds = 1 minute
360 seconds = 360 ÷ 60 = 6 minutes

```
       6
60)360
   -360
      0
```

 **EXERCISES**

1. Convert to minutes.

i) 13 hours   ii) 36 hours   iii) 45 hours

iv) 120 hours   v) 90 hours

2. Convert to hours.

i) 122 minutes   ii) 144 minutes   iii) 90 minutes

iv) 230 minutes   v) 78 minutes

3. Convert to seconds.

i) 32 minutes   ii) 54 minutes   iii) 72 minutes

iv) 120 minutes   v) 90 minutes

4. Convert seconds to minutes.

i) 60 seconds   ii) 86 seconds   iii) 450 seconds

iv) 320 seconds   v) 190 seconds

5. Write the time shown in the following clocks.

_____   _____   _____

_____   _____   _____

_____   _____   _____

6. Draw the two hands on the clocks showing the time given below.

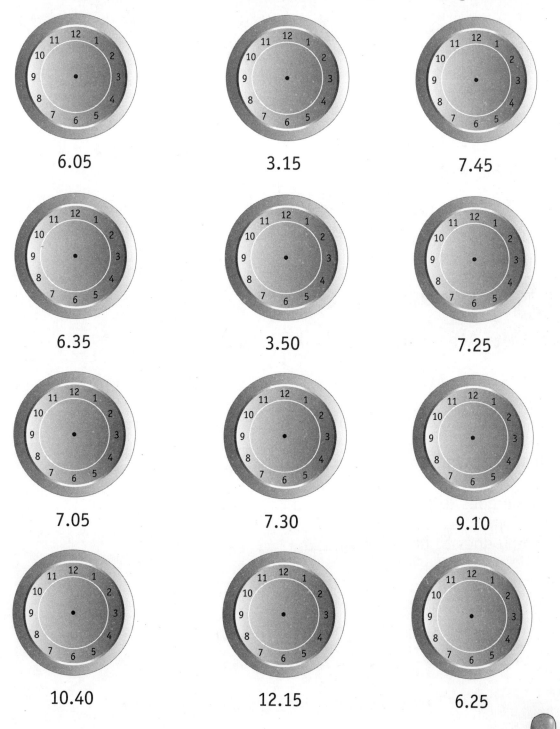

     5.30   2.10   3.55

AM and PM

The time from 12.00 midnight to 12.00 afternoon is denoted by AM (ante meridian) and the time from 12.00 noon to 12.00 midnight is denoted by PM (post meridian).

 EXERCISES

1. Write AM or PM in the blanks.

i) Ravi gets up at 6.30 _____

ii) His bus leaves for school at 7.20 _____

iii) Raju comes back from school at 2.30 _____

iv) He goes to play at 4.30 _____

v) Anu goes to bed at 9.30 _____

2. State the difference in terms of minutes between clock A and clock B.

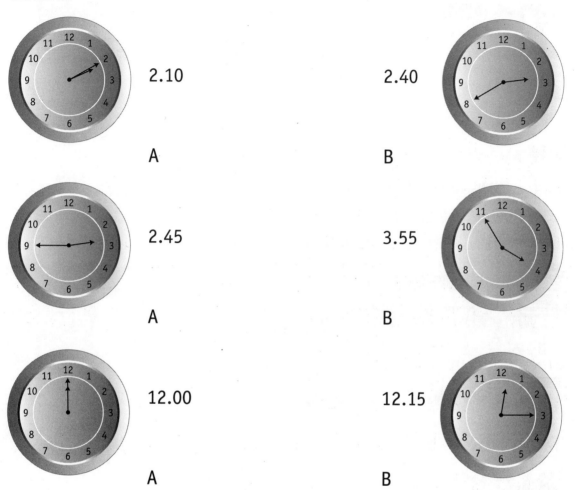

3. Neeta left home at 7.30 AM and reached school at 8.15 AM. How long did she take to reach school?

4. Ravi leaves office at 6.15 PM everyday and reaches home at 7.30 PM. How much time does he take to reach home?

HOURS AND MINUTES

Example

1. A school starts at 8.30 AM and closes at 1.10 PM. How long does the school work?

8.30 AM to 9 AM = 30 minutes
9.00 AM to 1 PM = 4 hours
1 PM to 1.10 PM = 10 minutes
Total time = 4 hours 40 minutes

2. A bus starts from Mumbai at 7.20 AM and reaches Pune at 11.45 AM. How long does the bus take to reach Pune?

7.20 AM to 8 AM = 40 minutes
8.00 AM to 11 AM = 3 hours
11 AM to 11.45 AM = 45 minutes
Total time = 3 hours 85 minutes = 4 hours 25 minutes

 ### EXERCISES

1. Ravi goes to office at 8.00 AM everyday and comes back home by 7.20 PM. How long is he out of his house?

2. A train leaves station A at 6.45 AM and reaches station B at 7.20 AM the next day. How long does the train take to travel from station A to station B?

3. An examination started at 11.45 AM and ended at 2.20 PM. What was the duration of the examination?

CALENDAR

Number of days between two dates

Example

1. How many days are there between 30 January and 14 February?

Days left in January = 31 - 29 = 2
Days in February = 14
Total number of days = 2 + 14 = 16
Therefore, there are 16 days between 30 January and 14 February

2. How many days are there between 13 June and 10 September?

Days left in June = 30 - 12 = 18
Days in July = 31
Days in August = 31
Days in September = 10
Total number of days = 18 + 31 + 31 + 10 = 90
Therefore, there are 90 days between 13 June and 10 September.

FINDING OUT THE DATE

1. What is the date 12 days after 23 March?

Days in March = 31 - 23 = 8
Days in April = 12 - 8 = 4
The date 12 days after 23 March is 4 April

EXERCISES

1. Tanay's vacation starts on 19th May and the school reopens on the 12th of July. How many days' holidays does he have?

2. An exhibition started on the 12th of September and ended on 30th September. How long did the exhibition last?

3. Tanuj leaves for a trip on the 15th of March. He takes 15 days to travel and come back home. On what date does he reach home?

Notes

Notes